AF771186

# ARTICULATION OF SOUND FORMS IN TIME

Some of these poems first appeared in *Conjunctions, Sulfur, Acts, Ironwood, Boundary, Luna Tack, Hambone,* and *Io.*

© Copyright Susan Howe, 1987.
ISBN 0-942433-11-4

# ARTICULATION OF SOUND FORMS IN TIME

Susan Howe

AWEDE
1987

from seaweed said nor repossess rest
scape esaid

*EXTRACT from a LETTER (dated June 8th, 1781,) of Stephen Williams to President Styles:*

"In looking over my papers I found a copy of a paper left by the Rev. Hope Atherton, the first minister of Hatfield, who was ordained May 10th, 1670. This Mr. Atherton went out with the forces (commanded by Capt. Turner, captain of the garrison soldiers, and Capt. Holyoke of the county militia) against the Indians at the falls above Deerfield, in May, 1676. In the fight, upon their retreat, Mr. Atherton was unhorsed and separated from the company, wandered in the woods some days and then got into Hadley, which is on the east side of the Connecticut River. But the fight was on the west side. Mr. Atherton gave account that he had offered to surrender himself to the enemy, but they would not receive him. Many people were not willing to give credit to this account, suggesting he was beside himself. This occasioned him to publish to his congregation and leave in writing the account I enclose to you. I had the paper from which this is copied, from his only son Jonathan Wells, Esq., who was in the fight and lived afterward at Deerfield and was intimately acquainted with the *Indians* after the war, did himself inform *me* that the *Indians* told *him* that after the fall fight, a little man with a black coat and without any hat, came toward them, but they were afraid and ran from him, thinking it was the Englishman's God, etc., etc.,"

*I*

## *Hope Atherton's Wanderings*

Prest try to set after grandmother
revived by and laid down left ly
little distant each other and fro
Saw digression hobbling driftwood
forage two rotted beans & etc.
Redy to faint slaughter story so
Gone and signal through deep water
Mr. Atherton's story Hope Atherton

Clog nutmeg abt noon
scraping cano muzzell
foot path sand and so
gravel rubbish vandal
horse flesh ryal tabl
sand enemys flood sun
Danielle Warnare Servt
Turner Falls Fight us
Next wearer April One

Soe young mayde in March or April laught
who was lapd M as big as any kerchief
as like tow and beg grew bone and bullet
Stopt when asleep so Steven boy companion
Or errant Socoquis if you love your lives
War closed after Clay Gully hobbling boy
laid no whining trace no footstep clue
"Deep water" he *must* have crossed over

Who was lapt R & soe grew bone bullet
as like tow and as another scittuation
Stopt when Worshp Steven boy companion
Abt noon and abt sun come Country Farm
Follow me save me thither this winter
Capt. Turner little horn of powder
Medfield Clay Gully hobbling boy
Sixteen trace no wanton footstep rest
Soe struck fire set the woods on fire

Two blew bird eggs plat
Habitants before dark
Little way went mistook awake
abt again Clay Gully
espied bounds to leop over
Selah cithera Opynne be
5 rails high houselot Cow
Kinsmen I pray you hasten
Furious Nipnet Ninep Ninap
little Pansett fence wth ditch
Clear stumps grubbing ploughing
Clearing the land

Antagonists lay level direction
Logic hail um bushell forty-seven
These letters copy for shoeing
was alarum by seaven bold some
Lady Ambushment signed three My
excuse haste Nipmunk to my loues
Dress for fast Stedyness and Sway
Shining at the site of Falls Jump
Habitants inning the corn & Jumps

Rash catastrophe deaf evening
Bonds loosd catcht sedge environ
Extinct ordr set tableaux
hay and insolent army
Shape of so many comfortless
And deep so deep as my narrative
our homely manner and Myself
Said "matah" and "chirah"
Pease of all sorts and best
courtesy in every place
Whereat laughing they went away

rest chondriacal lunacy

velc cello viable toil

quench conch uncannunc

drumm amonoosuck ythian

scow aback din

flicker skaeg ne

barge quagg peat

sieve catacomb

stint chisel sect

Otherworld light into fable
Best plays are secret plays

Mylord have maize meadow

have Capes Mylord to dim

barley Sion beaver Totem

W'ld bivouac by vineyard

Eagle aureole elses thend

Impulsion of a myth of beginning
The figure of a far-off Wanderer

Grail face of bronze or brass
Grass and weeds cover the face

Colonades of rigorous Americanism
Portents of lonely destructivism

Knowledge narrowly fixed knowledge
Whose bounds in theories slay

Talismanic stepping-stone children
brawl over pebble and shallow

Marching and counter marching
Danger of roaming the woods at random

Men whet their scythes go out to mow
Nets tackle weir birchbark

Mowing salt marshes and sedge meadows

Body perception thought of perceiving (half-thought

chaotic architect repudiate line Q confine lie link realm

circle a euclidean curtail theme theme toll function coda

severity whey crayon so distant grain scalp gnat carol

omen Cur cornice zed primitive shad sac stone fur bray

tub epoch too tall fum alter rude recess emblem sixty key

Epithets young in a box told as you fly

Posit gaze level diminish lamp and asleep(selv)cannot see

*is* notion most open apparition past Halo view border redden

possess remote so abstract life are lost spatio-temporal hum

Maoris empirical Kantian a little lesson concatenation up

tree fifty shower see step shot Immanence force to Mohegan

blue glare(essence)cow bed leg extinct draw scribe upside
even blue(A)ash-tree fleece comfort(B)draw scribe sideup

Posit gaze level diminish lamp and asleep(selv)cannot see

MoheganToForceImmanenceShotStepSeeShowerFiftyTree

UpConcatenationLessonLittleAKantianEmpiricalMaoris

HumTemporal-spatioLostAreLifeAbstractSoRemotePossess

ReddenBorderViewHaloPastApparitionOpenMostNotion *is*

blue glare(essence)cow bed leg extinct draw scribe     sideup
even blue(A)ash-tree fleece comfort(B)draw scribe     upside

Loving Friends and Kindred:—

When I look back

So short in charity and good works

We are a small remnant

of signal escapes wonderful in themselves

We march from our camp a little

and come home

Lost the beaten track and so

River section dark all this time

We must not worry

how few we are and fall from each other

More than language can express

Hope for the artist in America & etc

This is my birthday

These are the old home trees

2

## *Taking The Forest*

Corruptible first figure
Bright armies wolves warriors steers

scorned warning captive compulsion

Love leads to edge
Progress of self into illusion

Same and not the same
Cherubim intone their own litany

Universal separation
—Distant coherent rational system

Vault lines divergence
Atom keystone

Parmenides prohibition
End of passageway perceive surrounding

Consciousness grasps its subject
Stumbling phenomenology

infinite miscalculation of history

Great men thicker than their stories
sitting and standing

to mark suns rising and setting
Ridges of sand rising on one another

Mathematics of continua

fathomless infinitesimal fraction
sabbatical safety beyond seven

Empty arms cloud counterfeit

antecedent terror stretched to a whisper

Certain autumnal story

absent humanity
Mothers from their windows look

Forth now
Sunbeam

oakleaf wreathe possible symbol

splitting nature's shadow
splitting the world

hammer at nature to temper Terrible

Range loom and lacing Clear
companion

shuttle face lost     Prelude
Straw portion

seed to be called Sarah's laughter

Dodge tinder huge hush
dodge thoroughfare outside since

Heard trace tidings

Complexity kissing sense into empty

Thread gone
Tongue collect song

Logical determination of position

Scuttle and striving

Being into never ending sequence of

Becoming
Coming home through past ages

visible light on the legionaires' banner
with welcoming welcoming

and undersweeping

Double penetrable foreign sequel
By face to know helm

Prey to destroy in dark theme
Emblem offictitious narrative

Step and system

Collision and impulsion
Asides and reminders to myself

Lives to be seen pressing and alien

Fix fleeting communication

Carried away before a pursuer
Demonstration in a string of definitions

To walk a little
Night's kingdom lamentation dimming

Snatched idea
Recollection fallen away from ruin

Slide into elect unalterable

slipping from known to utmost bound

Paper plague odd ends reap

Faint legend is day
Mahomet touched a flower with the tips

of his fingers
Face of the voice of speech

Flaming heart on the extended hand
Sky red and lowering

Mountains pitched over to westward

Light inaccessible as darkness
Irradiation of infinite

Principles judgements must connect
Soul lying awake in sleep

Parabolic scholar
Follower

Hail of witches on heath
Wars under the earth

Day cut off
Or summer's rose

Quicksand swallows the bottom

Leaf carried out of invisible hearing
Accoustic signature

prefiguration
Degree of weight and absolute light

Summer has come again
mute time remarked it

Deep generations of the old pass over
Surging and separating

Flame choristers shuddering flame
Self formed in line

Linear theme
prefiguration without premeditation

Vision sweeps away

Immeasurable distance farthest Hebrides

Caused the universe
Contained the Universe

Rugged solacing on surge sheltering

Invisible festival
halves drawn into a circle

feigned necessity vanished from memory

Hymn to the light
hymn and psalm to the light

Eight points of a compass

mosaic pattern eastward dispelled

Summoners slip into ritual
Planctus of Rachel

Shadowy figures of the Magi
elongated figures trooping in stone

Parallel column
Geographical assertion

Straw mother heaved higher and higher
Self ceded to sphere

Pull world around streets
Prosper our journey

Class structure belonging to shelter

migratory odd scrap trilogy

Laughter hallows the house in summer
Thin home hover in theme

Far away at a periphery

Celt heaven
Lugboat of children thousand in superstition

streak is the sky
sliver of shimmering

disputation in dominion beyond sovereign

Puck's face of earth
Sign and laughingstock Puck

Fable into fretted parable flew off
An old theme frolicking

Walk on the ramparts

headlong dwindled to span

Predecessor and definition
incoherent inaccessible muddled inaudible

Speech was a cry for action

erroneous proof
erroneous choice

axe-edge wording of each ramping lion
*whirlwind handwritten*

Commonwealths kindle and climb

Tender scudding visible
My body goes out alone as I came

nameless numberless no plan

Chaos cry car rion empirical proof proving
Summary succession of spectators twisted

away

We turn suddenly

Lords of the Lay

Letters sent out in crystalline purity

Muddled and ravelled

Sigh by see

Smoke faces separate

Lore and the like

Sucked into sleeping

—Hegelian becoming

—Hugolian memory

Patriarchal prophesy at heels of hope

Futurity—

Shouting an offering

Fleet messengers falter

Obedient children elder and ever

Lawless center

Scaffold places to sweep

*unfocused future*

Migratory path to massacre

Sharpshooters in history's apple-dark

Bound Cupid sea washed

Omen of stumbling
Great unknown captaincy

centuries roam audible silence
whissing days roam own key

Second time somewhere
seclusion in symbol sovereign

Memory mutinies contours cling

Mother my name
pin-eyed children

sundial lair from mortal fear

hooed Peru

Prisoned stone guardian
geurdon of resurrection

fictitious deeps

Cries open to the words inside them
Cries hurled through the Woods

Threadbare evergreen season
Mother and maiden

Singing into the draft

Keen woes centuries slacken
woe long wars endurance bear

In forest splinter companion

essential simplicity of Thought
wedged back playmate of Remote

Hares call on Pan
To rhyme with reason revels run

Untraceable wandering
the meaning of knowing

Poetical sea site state
abstract alien point

root casket tangled scrawl

Mistletoe arrow
ascetic hero-shadow

Shelter secret in heart
were wound drawn out

Rhyme of Heaven open

Collision with human protection

Left home to seek Lost

Pitchfork origin

tribunal of eternal revolution
tribunal of rigorous revaluation

Captive crowned tyrant deposed
Ego as captive thought

Conscience in ears too late

Father the law
Stamped hero-partner

pledge of creditor to debtor

Destiny of calamitous silence
Mouth condemning me to absence

——Uneasy antic alibi

dimmed dimmest world
final fertile mantle of family

Leap from scratch to ward off

Home in a human knowing

Stretched out at the thresh
of beginning

Sphere of sound

Body of articulation chattering

an Assassin
shabby halo-helmet

hideout haystack hunter chamois

History of seedling and seduction
Kinship of infinite separation

Sight of thought

Crooked erratic perception
shoal ruin abyssal veil veiling

Braggart expert
discourse on dice

Face to visible sense gathers moss

Freedom's dominion of possible

Ear to parable
lilies spin glory

Adamant glides architrave front

Path to blest vanishing
kindled oracle vanishing

Stripped of metaphysical proof
Stoop to gather chaff

Face to fringe of itself
forseen form from far off

Homeward hollow zodiac core

omen cold path to goal

End of the world as trial or possible
trail

The Now that is Night
Time comprehended in Thought

Sullen chill uncertain
Solitude and chill uncertain

Glacier cloud drifting

nimbus of extinction
mimic tracery mimic swaddling

week of dull day under hell-sky

All things double on one another
On to pure purpose

Spinoza the lense grinder
Lenses and language

total systemic circular knowledge
System impossible in time

truant freedom of dream

Wishes fly off
Tender consecration of garden

frost ends the rose
november affirmation in negation

Shear against easternmost

eternal Ideal sequence
Out among haphazard children

sunny investigations of Permanence

Philology heaped in thin

hearing
Cult code mediation

(Cult cairn)
Symbol alloted to ocean

Talker
and intellectual attacker

edging and dodging

Outer shallow mute

Mackerel sky
wind-ripples

Legend of the King's hill

name of wildflowers
roost in neighboring

Scuffle eminence peacock keeper

Sign of sound
sibilant wind

Scanned chronicles clasp edges

rag veracity
ring of our bodies

Names are bridges to coast

permanence

Naked figure moving in color all flower

Occult ferocity of origin

each winged ambition
sand track wind scatter

Inarticulate true meaning

lives beyond thought
linked from beginning

Pilings of thought under spoken

Physiognomy of Liberty

far friend forever Nestling
Forfeit mortality

Cycles snare mastery
headlong centuries cycles ensnare

Face answers to face
limit and quiet Limit

Field of vision and field of future

Shadowy Icarian figuration

Vision closes over vision
Standpoint melts into open

wanton meteor ensign streaming

Girl with forest shoulder
Girl stuttering out mask or trick

aria out of hearing

Sound through cult annunciation
sound through initiation Occult

Enunciate barbarous jargon
fluent language of fanaticism

Green tree of severance
Green tree girdled against splitting

Transmutation of murdered Totem

Foresters move before error
forgotten forgiven escaping conclusion

Oak and old hovel grow gossamer

Shoal kinsmen trespass Golden
Smoke splendor trespass

Symmetry carried from country
frail counterfeit well met

Lost among equivocations
Emancipator at empyrean center

Anarchy into named theory
Entangled obedience

muffled discourse from distance
mummy thread undertow slough

Eve of origin Embla the eve
soft origin vat and covert

Green hour avert grey future
Summer summon out-of-bound shelter

Hook intelligence quick dactyl

Bats glance through a wood
bond between mad and made

anonymous communities bond and free

Perception crumbles under character
Present past of immanent future

Recollection moves across meaning
Men shut their doors against setting

Flocks roost before dark
Coveys nestle and settle

Meditation of a world's vast Memory

Predominance pitched across history
Collision or collusion with history

Summary of fleeting summary
Pseudonym cast across empty

Peak proud heart

Majestic caparisoned cloud cumuli
East sweeps hewn flank

Scion on a ledge of Constitution
Wedged sequences of system

Causeway of faint famed city
Human ferocity

dim mirror Naught formula

archaic hallucinatory laughter

Kneel to intellect in our work
Chaos cast cold intellect back

Hemmed trammels of illusion
rooted to shatter random

Firstborn of Front-sea
milestone by name farewell

Milestones bewitched millstones
Sleep passage from Europe

Otherworld light into fable

Negative face of blank force

Winds naked as March
bend and blend to each other

Fledgling humming on pathless

Old Double and old beginning Vain
Covergesture

Latin ends and French begins

Golden page third voyage
Caravels bending to windward

Crows fly low and straggling
Civilizations stray into custom

Struts structure luminous region
Purpose or want of purpose

Part of each kingdom of Possession

Only conceived can be seen
Original inventors off Stray

Alone in deserts of Parchment
Theoreticians of the Modern

--emending annotating inventing
World as rigorously related System

Pagan worlds moving toward destruction

Visible surface of Discourse

Runes or allusion to runes
Tasks and turning flock

Evening red enough for chivalry

Algorithms bravadoes jetsam
All Wisdom's plethora pattern

paper anacoluthon and naked chalk

Luggage of the prairie
Wagons pegged to earth

Tyrannical avatars of consciousness
emblazoned in tent-stitch

Five senses of syntax

Dear Unconscious scatter syntax
Sythe mower surrender hereafter

Dear Cold cast violet coronal

World weary flesh by Flesh bygone
Bridegroom

Last line of blue hills

Lost fact dim outline

Little figure of mother

Moss pasture and wild trefoil
meadow-hay and timothy

She is and the way She was

Outline was a point chosen
Outskirts of ordinary

Weather in history and heaven

Skiff feather glide house

Face seen in a landscape once

To kin I call in the Iron-Woods
Turn I to dark Fells last alway

Theirs was an archheathen theme
Soon seen stumbled in lag Clock

Still we call bitterly bitterly
Stern norse terse ethical pathos

Archaic presentiment of rupture
Voicing desire no more from here

Far flung North Atlantic littorals

Lif sails off longing for life
Baldr soars on Alfather's path

Rubble couple on pedestal
Rubble couple Rhythm and Pedestal

Room of dim portraits here there
Wade waist deep maidsworn men

Crumbled masonry windswept hickory

PRINTED LETTERPRESS
AWEDE BRITA BERGLAND
BOX 376 WINDSOR VERMONT
05089 MARCH 1987
EDITION OF
1000

8
2875